OUR PLANET EARTH
Wetlands
by Rebecca Sabelko
BLASTOFF! READERS
3
BELLWETHER MEDIA • MINNEAPOLIS, MN

Blastoff! Readers are carefully developed by literacy experts to build reading stamina and move students toward fluency by combining standards-based content with developmentally appropriate text.

Level 1 provides the most support through repetition of high-frequency words, light text, predictable sentence patterns, and strong visual support.

Level 2 offers early readers a bit more challenge through varied sentences, increased text load, and text-supportive special features.

Level 3 advances early-fluent readers toward fluency through increased text load, less reliance on photos, advancing concepts, longer sentences, and more complex special features.

★ **Blastoff! Universe**

Reading Level

Grade K

Grades 1–3

Grade 4

This edition first published in 2022 by Bellwether Media, Inc.

Library of Congress Cataloging-in-Publication Data

Names: Sabelko, Rebecca, author.
Title: Wetlands / by Rebecca Sabelko.
Description: Minneapolis, MN : Bellwether Media, 2022. | Series: Blastoff! readers. Our planet earth | Includes bibliographical references and index. | Audience: Ages 5-8 | Audience: Grades 2-3 |
Summary: "Simple text and full-color photography introduce beginning readers to wetlands. Developed by literacy experts for students in kindergarten through third grade"-- Provided by publisher.
Identifiers: LCCN 2021045043 (print) | LCCN 2021045044 (ebook) | ISBN 9781644876084 (library binding) | ISBN 9781648346194 (ebook)
Subjects: LCSH: Wetlands--Juvenile literature.
Classification: LCC GB622 .S23 2022 (print) | LCC GB622 (ebook) | DDC 333.91/8--dc23
LC record available at https://lccn.loc.gov/2021045043
LC ebook record available at https://lccn.loc.gov/2021045044

Editor: Kieran Downs Designer: Laura Sowers

Printed in the United States of America, North Mankato, MN.

Table of Contents

What Are Wetlands?

Wetlands are land areas that are wet at least part of the year. They have watery soil. Some are covered in water.

Wetlands are often found near bodies of water. They also form when water slowly pushes up through the ground.

swamp

Different types of wetlands are found all over the world. High **tides** flow into **mangrove forests** along coasts. **Swamps** are filled with tall trees.

Marshes are like flooded grasslands. Mosses grow in **peatlands** located in cold areas.

Wetlands help the earth in many ways. They store and clean water. **Pollutants** sink into the ground as the water sits. They become trapped.

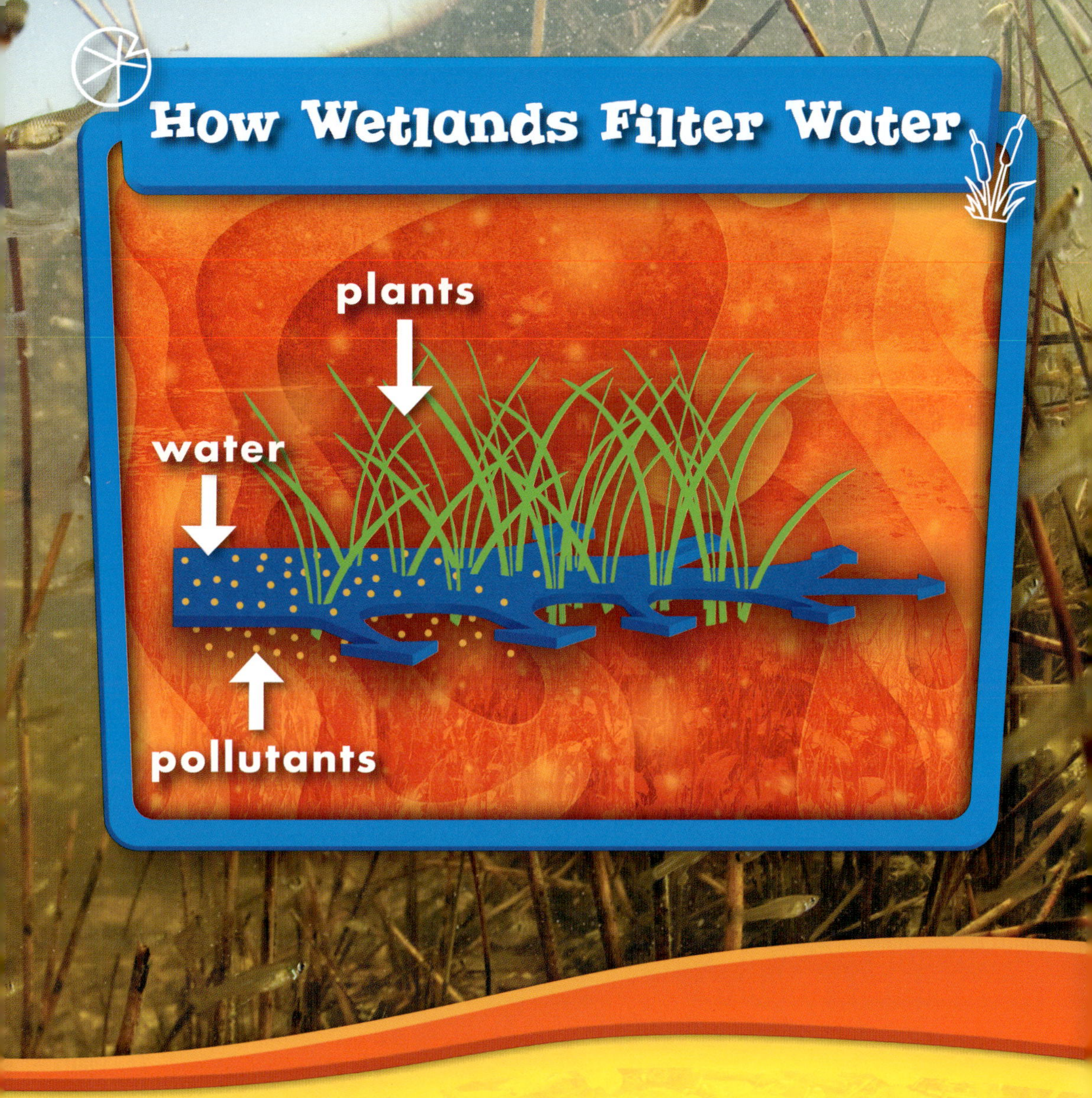

Wetland plants also **filter** pollutants. The plants help stop **erosion** of the land, too.

Wetlands also help during floods. They take in extra water when bodies of water overflow. They release it slowly to help lessen damage.

The Everglades

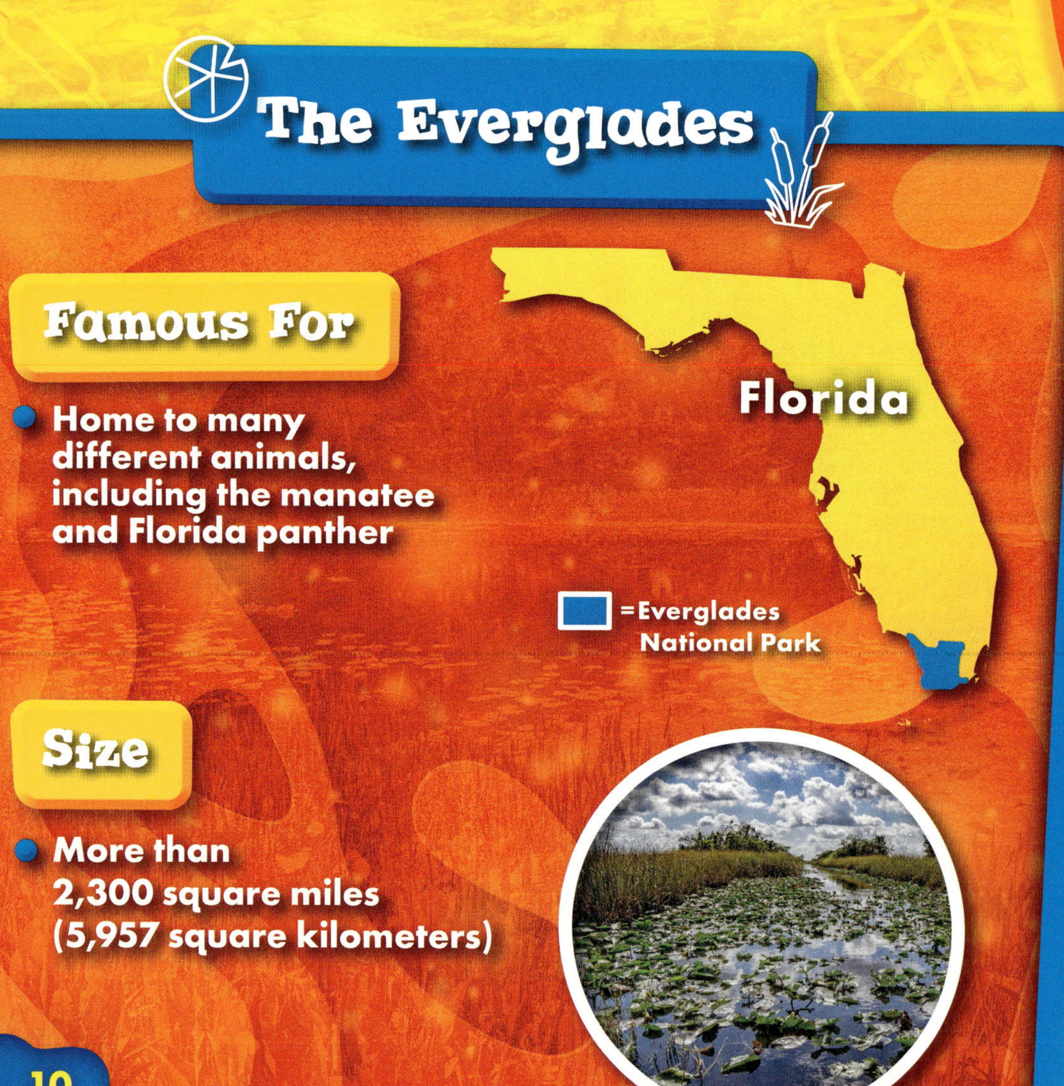

Famous For

- Home to many different animals, including the manatee and Florida panther

Size

- More than 2,300 square miles (5,957 square kilometers)

mangrove forest

Wetlands protect coasts, too. Mangrove forests help slow flooding and erosion during storms.

Plants and Animals

European golden plover

Many plants and animals depend on wetlands. Young perch swim in floodwaters. Plovers rest during **migrations**.

Frogs hide from snakes amongst floating plants. Bats swoop after mosquitoes. Otters feed on crayfish.

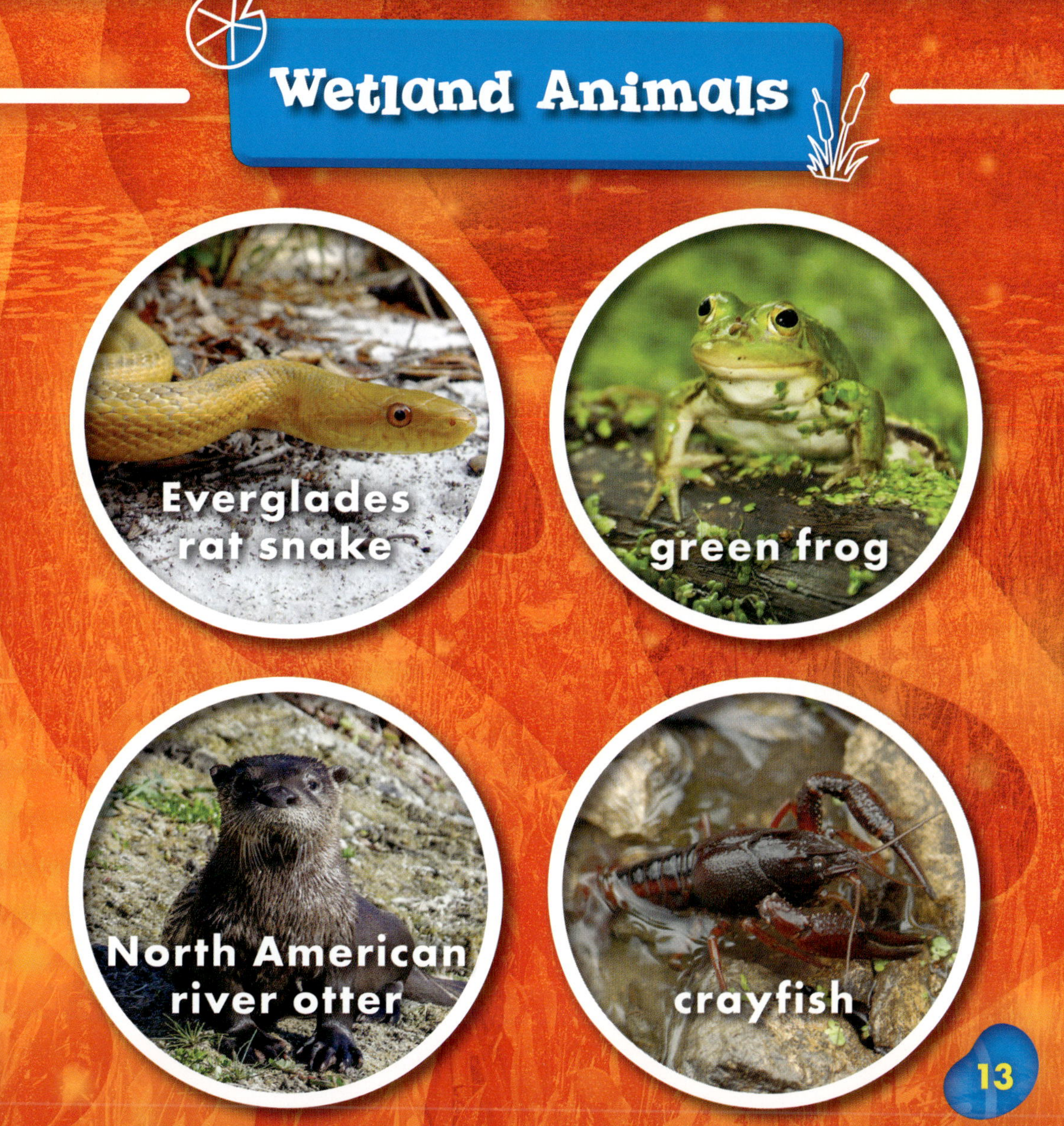

Some plants are only found in wetlands. Cattails sway in the breeze. Water lilies bloom. Cypress trees stand tall in swamps.

Lichens grow in cold wetlands. Dead plants slowly break down to form peat in these regions.

People and Wetlands

People use wetlands in many ways. They use the water for drinking. Many foods grow in these areas.

People enjoy outdoor sports in wetlands. Bird-watching is popular, too.

pollution

People destroyed wetlands to create cities and farms for many years.

Today, farms pollute the water. **Invasive species** harm other plants and animals. Rising sea levels destroy coastal wetlands.

How People Affect Wetlands

- Farms pollute wetland water

- Invasive species harm other plants and animals

- Rising sea levels destroy coastal wetlands

People began to understand the importance of wetlands in the 1970s. Many are protected today. But more needs to be done.

We must continue to learn about wetlands to keep them safe!

Glossary

erosion—the wearing away of land by the action of water, wind, or ice

filter—to remove

invasive species—a species that is not originally from a region that causes harm to its new environment

lichens—plantlike living things that grow on rocks and trees

mangrove forests—thick, tropical forests that can grow along coasts in salty swamp water

marshes—wetlands that are filled with grasses

migrations—trips from one place to another, often with the seasons

peatlands—wetlands that contain peat; peat is a dark material that forms from plants that partly decay in water.

pollutants—substances that make the earth dirty or unsafe

swamps—wetlands filled with trees and other woody plants

tides—the rise and fall of the oceans that occurs twice a day

To Learn More

AT THE LIBRARY

Nelson, Penelope S. *Everglades National Park.* Minneapolis, Minn.: Jump!, 2020.

Shea, Therese M. *20 Fun Facts About Wetland Habitats.* New York, N.Y.: Gareth Stevens Publishing, 2022.

Willis, John. *Wetlands.* New York, N.Y.: AV2, 2021.

ON THE WEB

FACTSURFER

Factsurfer.com gives you a safe, fun way to find more information.

1. Go to www.factsurfer.com.
2. Enter "wetlands" into the search box and click 🔍.
3. Select your book cover to see a list of related content.

Index

The images in this book are reproduced through the courtesy of: Irina Wilhauk, front cover; Africa Studio, p. 3; Bob Pool, pp. 4-5; Jon C. Beverly, p. 5; lazyllama, pp. 6-7; Andre Maceira, p. 7; komkrit Preechachanwate, p. 8; Michael Patrick O'Neill / Alamy Stock Photo, pp. 8-9; Andrea Izzotti, p. 10; Mazur Travel, pp. 10-11; Petr Simon, p. 12; Matt Jeppson, p. 13 (Everglades rat snake); RafalKwiatkowski, p. 13 (green frog); Microfile.org, p. 13 (North American river otter); Rusty Dodson, p. 13 (crayfish); Zelijko Radojko, p. 14; tonanakan, pp. 14-15; Leecy Jones, p. 15; Dirk M. de Boer, p. 16; Tompix, pp. 16-17; jaimie tuchman, p. 18; Shuang Li, pp. 18-19; higrace, pp. 20-21; Hintau Aliaksei, p. 23.